CONSTRUCTION INDUSTRY TRAINING

DOMESTIC GAS SERVICES

GAS SAFETY REGULATIONS

Revised and Reprinted 1991

©Construction Industry Training Board, 1987   ME 202/1

Published by:
Construction Industry Training Board
Third Edition, 1991
ISBN 0 902029 84 3
©Construction Industry Training Board, 1987

CONTENTS

1. **Legal Installation Requirements**

Gas Acts	1
Building Acts	5
Electricity Acts and Regulations	6
Water Acts	7

   Other Relevant Legislation

Health and Safety at Work Act	9
Construction Regulations	10
Construction (Health and Welfare) Regulations	10
Protection of Eyes Regulations	11
Health and Safety (First Aid) Regulations	11
Control of Substances Hazardous to Health Regulations	11
Liquefied Petroleum Gas	11

2. **The Gas Safety (Installation and Use) Regulations**

Notes	13
Arrangement of Regulations	14
Part A: General	16
Part B: Gas Fittings - General Provisions	18
Part C: Meter Installations	25
Part D: Installation Pipes	31
Part E: Gas Appliances	38
Part F: Miscellaneous	45
Explanatory Note	48

Appendix A: BSI Codes of Practice for Installing Domestic Gas Appliances	51

ACKNOWLEDGEMENT

The CITB wish to thank the following organisations for their assistance in the development of the Gas Services Training Courses.

    CORGI
    Skills Training Agency
    Gas Teachers Association

GAS SERVICES TRAINING COURSE

1. LEGAL INSTALLATION REQUIREMENTS (GAS APPLIANCES)

Introduction

The installation of a gas appliance must comply with all statutory requirements (including any local by-laws) and such other regulations as may be relevant.

Many gas installations are connected to electricity and/or water supplies. The installer of a gas appliance needs to be aware of the legal installation requirements for 'Gas', 'Electricity' and 'Water'.

The principal statutory and other requirements to be observed are those made under the Gas, Electricity and Water Acts, together with the London Building Acts, Public Health Acts, Clean Air Act, Fire Precautions Act, Building (Scotland) Act and the Health and Safety at Work Etc. Act. Regulations and by-laws made under these Acts which affect the installation of gas appliances are described under separate headings as follows:-

- Gas Acts
- Building Acts
- Electricity Acts
- Water Acts

GAS ACTS

The **Gas Act 1948** (which applied throughout Britain) made provision for safety regulations to be made by the Secretary of State for Trade and Industry. These came into operation on the 1st December 1972 and were entitled **The Gas Safety Regulations 1972**.

Although the Gas Act 1948 was repealed by the Gas Act 1972 which transferred responsibility for gas safety to the Secretary of State for Energy, the Gas Safety Regulations 1972 continued in force as if made under the Gas Act 1972.

The Gas Safety Regulations 1972 were administered by the Department of Energy (Gas Standard Branch) and applied throughout Britain, to gas supplied by British Gas. Installation work, service work and the use of gas by customers had to conform to these requirements. Any person contravening the Regulations could be fined.

The Gas Safety Regulations 1972 were supplemented by **The Gas Safety (Rights of Entry) Regulations 1976**, which came into operation on the 7th March 1977 and gave any officer authorised by the British Gas Corporation rights of entry, disconnection and discontinuance of supply.

The 1972 Gas Act was repealed and re-enacted by the **Oil and Gas (Enterprise) Act 1982** and amended by the Criminal Justice Act 1982.

The Gas Safety (Rights of Entry) Regulations 1983 superseded the Gas Safety (Rights of Entry) Regulations 1976 and came into operation on the 21st November 1983. They gave any officer authorised by *the relevent authority* rights of entry, disconnection and discontinuance of supply.

NOTE: The expression 'the relevant authority' is defined by section 31(8) of the 1972 Act-

> *(a) in relation to dangers arising from the distribution of gas by the British Gas Corporation or from the use of gas supplied by the Corporation, as meaning the Corporation, and*
>
> *(b) in relation to dangers from the distribution of gas by a person other than the Corporation, or from the use of gas supplied by such a person, as meaning the Secretary of State.*

Gas safety responsibilities were transferred from the Secretary of State for Energy to the Secretary for Employment on February 1st 1984.

The Secretary of State, in exercising the powers conferred on him, laid before Parliament new Regulations (on the 6th September 1984) revising legislation on the safe use of mains gas and the installation of appliances. These are **'The Gas Safety (Installation and Use) Regulations 1984**.

The new Regulations revoked Parts III to VI of the Gas Safety Regulations 1972; (the remainder of the Gas Safety Regulations 1972 subject to some amendments, remained in force). The new Regulations also amended the Gas Safety (Rights of Entry) Regulations 1983 by increasing the maximum penalty for breach of both these Regulations to £2,000.

Further amendments were made to the Gas Safety (Installation and Use) Regulations 1984 on March 30th 1990. These amendments come into force on March 30th 1991.

The Gas Safety (Installation and Use) Regulations 1984 (Amended 1990)

The Regulations are concerned with work on gas fittings where gas is supplied to premises through pipes. Work on gas fittings covers just about any action taken concerning gas, valves, meters, governors, appliances and other fittings.

The amendment prohibits an employer from allowing his employees to carry out work in relation to gas fittings unless the employer is a member of a class of persons approved by the Health and Safety Executive. This same ruling applies also to self-employed persons.

The amendment also states that 'no person shall carry out any work in relation to a gas fitting unless he is competent to do so'.

The Regulations require gas fittings to be installed and used in such a way that the public are protected, as far as is practicable, from personal injury, fire, explosion or other damage arising from the use of gas **supplied through pipes**. They cover all premises except for mines and factories (which are covered by other legislation). The Regulations apply to all use of gas, whether the gas is supplied by the British Gas Corporation or *other supplier.*

The Regulations place obligations on four categories of people:-

- the supplier of gas including, in some cases, landlords supplying gas to their tenants.

- employers (this includes the self-employed)

- those carrying out work on gas fittings, including not just installers but those servicing gas appliances.

- those responsible for premises, (who will usually be the owner or occupier, but could be, for example, the manager of a shop) and persons using gas.

The Regulations are enforced by the Health and Safety Commission and Health and Safety Executive.

Obligations on **gas suppliers** include the following:

- emergency controls shall be provided (8(1))

- emergency notices on meters shall be provided (14)

- disused but still live services shall be marked (15(3)(a)(ii)) and finally disconnected at the main (15(3)(b))

- if work is carried out on a pipe, or an appliance is installed at a time when gas is not being supplied to the premises, the supplier must ensure that the pipe is purged or the appliance tested before supplies commence (21(3) and 33(3)).

Obligations on **employers** including the **self-employed** include the following:-

- must ensure any person carrying out work to a gas fitting is competent (3(1))

- must be a member of a class of persons approved by the Health and Safety Executive (3(3)) (eg. CORGI registered)

Obligations on **those carrying out work** include the following:-

- they must be competent
- the gas fittings which they install must be sound (Regulations 4(1))
- fittings must be protected against undue risk of damage (6(1))
- emergency controls must work and be marked in a specified manner (8(2) and (3))
- meters must not be placed in certain places unless they are protected against fire (11(1) - (3))
- primary meters must be protected by gas governors (13)
- pipes must be protected against failure by movement (18(1))
  gas pipes in 'common parts' of buildings must be clearly marked (22(1))
- appliances must be in a safe condition before they can be installed (25(1)), be readily accessible for maintenance (28) and tested (33).
- flues and ventilation must be suitable for the appliance (25 and 27)
- fittings must not be left unattended (5(2))
- no source of ignition must be used when gasways are exposed (5(4))
- pipes shall be tested after being worked on (21(1))

Obligations on **persons responsible for premises** include the following:

- ensuring that any electrical cross-bonding is carried out if an installation pipe has been connected to a meter (17(2))
- ensuring that marked pipes (see (g) remain so marked (22)(2))
- take all reasonable steps to shut off the supply of gas if an escape of gas in the premises is brought to his or her notice (35(1)) and if gas continues to escape, to inform the gas supplier.

The **Gas Act 1986** provides for the privatisation of the **British Gas Corporation** and the formation of **British Gas PLC**. It replaces sections of the Gas Act 1972 (as amended by the Oil and Gas (Enterprise) Act 1982) and makes provisions for public gas suppliers and others to be authorised to supply gas through pipes.

BUILDING ACTS

Other statutory requirements must also be complied with. These requirements differ, depending where in Britain the appliance is installed. Regulation 26 indicates that a gas appliance must be installed in accordance with the requirements of building regulations and by-laws which apply in Great Britain.

These include:-

In the **Inner London** area:

The Building (Inner London) Regulations 1987

London Building Acts 1930

London Building Acts (Amendment) 1939

In **England and Wales** (excluding the Inner London area):

The Building Regulation (Amendment) Regulations 1989

In **Scotland**:

The Building Standards (Scotland) Regulations

The Building (Inner London) Regulations 1987

Under these regulations, building control procedures in Inner London became the same as in England and Wales. All the technical requirements of the Building Regulations 1985 and the later Building Regulations (Amendment) 1989 now apply in place of the Construction By-laws made under the previous London Building Acts. However certain additional requirements special to Inner London made under the London Building Act 1930 and the London Building Acts (Amendments) still apply.

The Building Regulations (Amendment) Regulations 1989

These regulations include amendments to the **Building Regulations 1985**, and came into force on 1st April 1990. The amendments affected Approved Document "J" which covers heat producing appliances including gas appliances as follows:-

— J1 AIR SUPPLY. Requiring an installation to receive sufficient air for the propel combustion of the level and operation of the fuel.

— J2 DISCHARGE OF PRODUCTS OF COMBUSTION. Requiring an installation to be capable of normal operation without the products of combustion becoming a hazard to health.

— J3 PROTECTION OF BUILDING. Where normal operation without causing damage by heat or fire to the fabric of the building is a requirement.

The Building Standards (Scotland) 1990

Previous regulations have been superseded.

The Building Standards (Scotland) 1990 are administered by the Scottish Development Department, and like the preceding building standards regulations contain requirements relating to the installation of gas appliances.

The majority of these requirements are given in Part F (Chimneys, flues, hearths and the installation of heat producing appliances).

NB. The current Building Regulations should always be used.

ELECTRICITY ACTS

The Electric Lighting (Clauses) Act 1899, the Electricity (Supply) Acts 1882 - 1936 allowed provision for supply regulations to be made by the Electricity Commission. These regulations came into operation on 1st January 1937, and were **The Electricity Supply Regulations 1937**. The current act is **The Electricity Supply Regulations 1988**.

The other statutory regulation to be complied with is **The Electricity at Work Regulations 1989** which came into force on 1st April 1990.

These Regulations apply to all electrical equipment and systems and require that such installations must not give rise to danger; should be suitably insulated and protected and provide for the installation to be isolated, or cut off, or the current reduced in the event of a fault. The Regulations also require equipment and installations to be properly identified and labelled.

It is now illegal to work on live electrical systems unless there is no other way in which work can be done.

The Regulations create duties for employers, the self-employed and employees and cover all aspects of electrical work, requiring that persons who work with electricity are competent. To be considered competent, a person must have:

- adequate knowledge of electricity
- good experience of electrical work
- an understanding of the system being worked on
- practical experience of that type of system
- knowledge of the hazards that might arise and the precautions that need to be taken
- the ability to immediately recognise unsafe situations

IEE Wiring Regulations

The **Regulations for Electrical Installations (Fifteenth Edition)** (referred to as the IEE REGULATIONS) are not statutory regulations, except in Scotland. They are issued by the Institution of Electrical Engineers and are designed to provide for the safety of electrical installations in and about buildings generally. Compliance with the IEE Regulations will, in general, satisfy the requirements of the Electricity Supply Regulations, and of the Building Standards (Scotland) Regulations. The IEE Regulations will also satisfy the requirements of the statutes covering electrical installations in factories, cinemas, mines and quarries.

The fifteenth edition of these Regulations is shortly to be amended. It is important that reference is always made to the current edition of the Regulations.

WATER ACTS

The Water Act 1945 applies in **England and Wales** allowing provisions for Water By-Laws to be made by statutory water undertakings throughout England and Wales for preventing waste, undue consumption, misuse or contamination of water supplied by them. Model Water By-Laws, were issued, on which the statutory water undertakings base their own Water By-Laws.

The Water Act 1973 came into effect on the 1st April 1974 and led to a reorganisation of the water industry in England and Wales, and the creation of ten new Regional Water Authorities, nine in England and one in Wales. As a consequence they became responsible for the management of water resources.

The Water Act 1989 provides for the establishment of a National Rivers Authority, and a Director General of Water Services. It also provides for regulations to be made in respect of the appointed companies, following the public sale of shares in the regional water authorities and water undertakings.

It should be remembered that it is the current Model Water By-Laws that must be taken into account when installing gas appliances using water.

The latest water Act to come into effect in Scotland is the Scotland Water Act 1980; however whilst this legislation remains in effect, it is the new Local Water By-Laws (replacing the Model Water By-Laws 1986) that need to be considered when installing gas appliances. These By-Laws are similar to the Model Water By-Laws taking effect in England and Wales.

Any person contravening the Water By-Laws, which are administered by the water authorities, can be fined.

Differences occurring in the requirements of the Water By-Laws in Britain can to a great extent be attributed to the fact that the constituents of water and subsoil vary from area to area. For example in areas where the water supplied is likely to cause de-zincification of copper alloy pipes and fittings, the Water By-Laws would prohibit the use of such pipes and fittings unless made from a copper alloy which was resistant to de-zincification, (eg. gun metal or alpha-brass of a type inhibited against de-zincification).

Summary

The legislative and other regulatory requirements which must be taken into account when installing or working on gas appliances are summarised below.

(i) In **Inner London**:

- Gas Safety (Installation and Use) Regulations 1984
- Gas Safety (Installation and Use) (Amendment) Regulations 1990
- The Building (Inner London) Regulations 1987
- London Building Acts 1930
- London Building Acts (Amendments) 1939
- Electricity at Work Regulations 1989
- IEE Regulations – Current Edition
- Model Water By-Laws 1986

(ii) In **England and Wales**:

- Gas Safety (Installation and Use) Regulations 1984
- Gas Safety (Installation and Use) (Amendment) Regulations 1990
- Building Regulations (Amendment) Regulations 1989
- Electricity at Work Regulations 1989
- IEE Regulations – Current Edition
- Model Water By-Laws 1986

(iii) In **Scotland**:

- Gas Safety (Installation and Use) Regulations 1984
- Gas Safety (Installation and Use) (Amendment) Regulations 1990
- The Building Standards (Scotland) 1990
- Electricity at Work Regulations 1989
- IEE Regulations – Current Edition
- Local Water By-Laws 1989

In all cases there may also be local by-laws which apply to the installation of gas appliances.

OTHER RELEVANT LEGISLATION

Health and Safety at Work etc. Act 1974

This Act provides a comprehensive legislative framework to promote, stimulate and encourage high standards of health and safety at the workplace. Its ultimate aim is to promote safety awareness and effective safety standards in every organisation.

One of the main aims of the Act is to involve **everyone** – management, the employees, the self-employed, the employees' representatives, the controllers of premises and the manufacturers of plant, equipment and materials – in matters of health and safety. The Act also deals with the protection of the public, where they may be affected by the activities of people at work.

The Act is an enabling measure, superimposed on existing health and safety legislation, (eg. the Factories Act or the Construction Regulations). This earlier legislation will remain in force until it is progressively repealed or replaced by improved and revised Regulations and Approved Codes of Practice issued under the Health and Safety at Work Act.

Regulations made under the Act impose a statutory duty and will be supplemented by Approved Codes of Pratice which have a special legal status. These are not statutory requirements, but may be used as evidence that statutory requirements have been contravened in criminal proceedings.

Regulations and Approved Codes of Practice may be made under a whole variety of health and safety topics, ranging from specific hazards to training.

General fire precautions at most places of work can now be dealt with by the fire authorities, eg. general precautions and means of escape in most factories, offices, shops etc. The Commission, remain responsible for control over 'process risks', that is, risks of outbreak of fire associated with particular processes or particular substances.

The Health and Safety at Work Act 1974 is dealt with in greater detail in the CITB Domestic Gas Services publication **Safety at Work** (ME 202/2).

Health and Safety Commission

Part 1 of the Act establishes the Health and Safety Commission comprising representatives of both sides of industry and the local authorities. It has responsibility for developing policies in the health and safety field.

The Commission issue Approved Codes of Practice and sometimes Guidance Notes (of a purely advisory nature) to support these.

Health and Safety Executive

The Health and Safety Executive is a separate statutory body appointed by the Commission which works in accordance with directions and guidance from the Commission. The Executive enforce legal requirements, as well as providing an advisory service to both sides of industry. The old Inspectorates, (eg. Factory Inspectorate) in the health and safety field have been brought within the Executive instead of working independently within several Government Departments.

The Construction Regulations

The Construction (General Provisions) Regulations, 1981 are concerned with the protection of those employed in building and construction work. They include the use of mechanically propelled vehicles and trailers (Reg 34): guards and fences for machinery (Reg 42): the use of electricity on site (Reg 44): protection from falling materials (Reg 46): the lighting of working places (Reg 47): lifting excessive weights (Reg 53).

The Construction (Lifting Operations) Regulations, 1981. As the title implies, these regulations are mainly concerned with the use of lifting appliances, cranes, derricks, winches etc. but include the use of simple gin wheels and block and tackle. They contain regulations regarding the inspection and testing of equipment, including chains, ropes and slings, safe working loads and the keeping of records.

The Construction (Working Places) Regulations, 1966 are mainly concerned with scaffolds and working platforms. They include provision of safe means of access, (which includes ladders, stairs, gangways etc.), and the safety of working platforms (Reg 6): ladders used to support a platform (Reg 14): trestle platforms (Reg 21): the maintenance and use of ladders and folding step-ladders (Reg 31, 32) and work on roofs and at height. They also confirm that it is the duty of an employer to ensure that scaffolds, ladders and working platforms etc, used by his employee are safe, even though they may be owned or erected by another (Reg 23).

The Construction (Health and Welfare) Regulations, 1966

Impose requirements for the health and welfare of persons employed at places where building operations and works of engineering construction are carried on. The Regulations contain provisions as to shelter from the weather, accommodation for clothing and for taking of meals, protective clothing, washing facilities and sanitary conveniences.

The Construction (Head Protection) Regulations 1989

These Regulations require the wearing of suitable head protection on all building and construction sites unless there is no risk of head injury other than the person falling. The Regulations cover "building operations" as well as "engineering construction".

The Protection of Eyes Regulations, 1974

The Protection of Eyes Regulations make provision for the protection of eyes of employees engaged in specified processes, and also persons not so employed, who may be at risk. Many of the specified processes are likely to be used in construction operations.

The protection provided should conform to specifications approved by the Chief Inspector of Factories and, as listed in the Regulations, includes goggles, visors, spectacles and face screens. It also includes fixed shields, either free standing or attached to machinery or plant.

Health and Safety (First Aid) Regulations, 1981

The Regulations provide a flexible framework within which individual undertakings can develop effective first aid arrangements appropriate to their workplace and size of work force. First aid requirements under earlier legislation* have been repealed or revoked in Schedules 1 and 2 of the Regulations.

* The Construction (Health and Welfare) Regulations 1966. Regulations 3(2), 4(2), 5 to 10. The Schedule.

Control of Substances Hazardous to Health Regulations 1988

These Regulations (usually referred to as COSHH) contain statutory duties for employers, employees, and self employed persons, including contractors and sub-contractors. The aim of the regulations is the protection of employees and others from the effects of working with substances hazardous to health.

The COSHH Regulations require you to identify and control hazardous substances in use at work.

Liquefied Petroleum Gas (LPG)

Liquefied Petroleum Gas (LPG) in the form of Butane and Propane is widely used in construction site processes and operations, cooking, heating and lighting. The principle hazard associated with LPG is **fire**. It is essential that precautions are taken to limit the risk involved.

The principal legislation concerning LPG is as follows:

- Highly Flammable Liquids and Liquefied Petroleum Gases Regulations 1972

    These regulations apply to places to which the Factories Act 1961 applies (factories, construction sites, warehouses etc.) and impose requirements for the protection of employees.

- Health and Safety at Work etc Act 1974

    The Health and Safety at Work Act does not specifically mention LPG: nevertheless anybody involved with LPG should be aware of the general requirements of this Act and in particular, Sections 2, 3 and 6.

- The Gas Safety (Installation and Use) Regulations 1984
    The Gas Safety (Installation and Use) (Amendment) Regulations 1990

    These regulations are concerned with work on gas fittings in domestic and commercial premises, where the gas is supplied through pipes. They require gas fittings to be installed and used in such a way, that as far as is practicable, the public is protected. Persons carrying out work must be competent. Employers are responsible for ensuring that these regulations are complied with.

References should also be made to the following HSE Guidance Notes:

The keeping of LPG in cylinder and similar containers (CS 4)

The storage of LPG at fixed installations (CS 5)

Storage and use of LPG on construction sites (CS 6)

Other References

The Building Regulations 1985 (England and Wales) with amendments

London Building Acts (Inner London Borough)

The Building Standards (Scotland) Consolidations

Electrical Safety Code, The Institute of Petroleum

BS 5482: Part 1, 1979 (1989) Code of Practice for Domestic butane – and propane – gas burning installations; installations in permanent dwellings. Part 2, 1977 (1988) Installations in caravans and non-permanent dwellings. *Note: These British Standards were revised on the dates in the brackets.*

GAS SERVICES TRAINING COURSE

2. The GAS SAFETY (Installation and Use) Regulations 1984
The Gas Safety (Installation and Use) (Amendment) Regulations 1990

This publication, containing the Government's Gas Safety (Installation and Use) Regulations 1984, as Amended 1990 has been reproduced by kind permission of the Controller of HMSO.
The majority of these Regulations came into force on 24th November 1984, with the remainder – mostly concerning gas 'suppliers' responsibilities – coming into force on 24th February 1985. The Amendment Regulations came into force on March 30th 1991.

The Regulations are printed in Roman type. **Additional comment is printed in** *italic (Page 16 onwards).*

IMPORTANT NOTES

1. The Regulations now apply to any fuel gas supplied through pipes. All service work, installation work and the safe use of gas must conform to the requirements of the Regulations. Failure to comply with a Regulation could lead on summary conviction to a fine of up to £2,000.

2. The Regulations do not cover work carried out in factories or mines and quarries (as defined in the relevant Acts) but will be used by the Factories Inspectorate as a guide to good practice in such premises.

3. The essence of the Regulations is safety and compliance should be ensured by following recognised good practice for installation and servicing work.

4. The final and authoritative interpretation of these Regulations is a function of Courts of Law only. CITB therefore accepts no liability for loss, damage, or injury which arises out of the use of or reliance upon this guide or any part there of.

5. The scope of these Regulations has been widened to include *any* gas supplied through 'pipes' and used for lighting, heating or motive power.

6. The Gas Safety Regulations 1972 have not been repealed but amended and now cover service pipes only.

7. The Regulations place requirements on gas suppliers, installers and responsible persons to provide information notices relating to the location of emergency control valves, etc., and other safety information.
Regulations 8(2)c, 8(3), 14(1), 14(2), 16,22(i) and 23(i)

8. No Regulation should be considered in isolation

STATUTORY INSTRUMENT 1984 No. 1358 : GAS and 1990 No. 824 : GAS

The Gas Safety (Installation and Use) Regulations 1984
The Gas Safety (Installation and Use) (Amendment) Regulations 1990

ARRANGEMENT OF REGULATIONS

PART A GENERAL

1. Citation and commencement.
2. General interpretation and application.

PART B GAS FITTINGS - GENERAL PROVISIONS

3. Qualification and supervision.
4. Materials and workmanship.
5. General safety precautions
6. Protection against damage.
7. Existing gas fittings.
8. Emergency controls
9. Electrical continuity - general

PART C METER INSTALLATIONS

10. Interpretation of Part C.
11. Meters - general provisions
12. Meter boxes.
13. Governors.
14. Meters - emergency notices.
15. Primary meters.
16. Secondary meters.

PART D INSTALLATION PIPES

17. Safe use of pipes.
18. Enclosed pipes.
19. Protection of buildings.
20. Clogging precautions.
21. Testing and purging of pipes.
22. Marking of pipes.
23. Large consumers.

PART E GAS APPLIANCES

24. Interpretation of Part E
25. Gas appliances - safety precautions.
26. Building legislation.
27. Flues.
28. Access.
29. Manufacturer's instructions.
30. Room-sealed appliances.
31. Suspended appliances.
32. Flue dampers.
33. Testing of appliances.
34. Unsafe appliances.

PART F MISCELLANEOUS

35. Escape of gas.
36. Penalty.
37. Exception as to liability.
38. Amendment of Gas Safety Regulations 1972 and Gas Safety (Rights of Entry) Regulations 1983.

The Secretary of State, in exercise of the powers conferred on him by sections 31(1), 42(2) and 45(3) of the Gas Act 1972 (a) and of all other powers enabling him in that behalf, hereby makes the following Regulations:-

---

(a) 1972 c.60; section 31 was repealed and substituted by section 14 of the Oil and Gas (Enterprise) Act 1982 (c.23); section 42(2) was amended by sections 40, 46 and 54 of the Criminal Justice Act 1982 (c.48).

PART A: GENERAL

Citation and commencement *(1990 Amendment)*

1. (1) These Regulations may be cited as the Gas Safety (Installation and Use) (Amendment) Regulations 1990 and shall come into force on 30th March 1991.

General interpretation and application

2. (1) In these Regulations -

    'emergency control' means a valve for shutting off the supply of gas in an emergency;

    'flue' means a passage for conveying the products of combustion from a gas appliance to the external air and includes any part of the passage in a gas appliance ventilation duct which serves the purpose of a flue;

    'gas appliance' means an appliance designed for use by a consumer of gas for lighting, heating, motive power or other purposes for which gas can be used;

    'installation pipe' means any pipe, not being a service pipe (other than any part of a service pipe comprised in a primary meter installation) or a pipe comprised in a gas appliance, for conveying gas for a particular consumer and any associated valve or other gas fitting,

    The gas carcass or supply pipes installed after the service valve. It excludes any pipe which forms part of a gas appliance;

    'meter by pass' means any pipe and other gas fittings used in connection with it through which gas can be conveyed from a service pipe to an installation pipe without passing through a meter;

    'primary meter' means a meter connected to a service pipe for ascertaining the quantity of gas supplied through that pipe;

    'primary meter installation' means a primary meter and the pipes and other gas fittings used in connection with it, including any meter bypass, installed between the outlet of any service valve, or, if there is no service valve, the outlet of the service pipe and the outlet connection of the meter or the outlet of the common connection of the meter and any meter bypass or any other primary meter as the case may be;

    'the responsible person', in relation to any premises, means the occupier of the premises or, where there is no occupier or the occupier is away, the owner of the premises or any person with authority for the time being to take appropriate action in relation to any gas fitting therein;

(a) S.I. 1972/1178, amended by S.I. 1976/1882, 1983/1575.

'service valve' means a valve for controlling a supply of gas incorporated in a service pipe and not situated inside a building;

Where there is a service valve in a service pipe, that part of the service pipe which is downstream of the service valve is part of the 'primary meter installation' and is a 'gas fitting' as well as an 'installation pipe';
the service valve could be:-

(a) meter control in an external meter box
(b) service entry tee for indoor meter installation
(c) below ground valve for schools, colleges, etc.

'work', in relation to a gas fitting, means work of any of the following kinds, that is to say -

(a) installing the fitting;

(b) maintaining, servicing, permanently adjusting, repairing, altering or renewing the fitting or purging it of air or gas;

(c) where the fitting is stationary, changing its position;

(d) removing the fitting.

For the first time the kinds of work covered by the Regulations are defined.

(2) For the purposes of these Regulations -

(a) the expression 'gas fitting' does not include any part of a service pipe except any part comprised in a primary meter installation;

This reflects the scope of these Regulations – 'service pipes' will be covered by other Regulations; in the interim, parts of the Gas Safety Regulations 1972 remain in force.

(b) any reference to installing a gas fitting includes a reference to converting any pipe, fitting, meter, apparatus or appliance to use gas supplied through pipes; and

(c) a person providing, for use in a flat or part of a building let by him, gas supplied to him shall not be in so doing deemed to be supplying gas.

The gas supplier to the primary meter (in flats or sub-let properties) is the 'supplier'. Landlords etc., providing gas to the secondary meters are not deemed to be the 'supplier'.

(3) Nothing in these Regulations shall apply in relation to the supply of gas to, or anything done in respect of a gas fitting at -

(a) a mine within the meaning of the Mines and Quarries Act 1954(a) or any place deemed to form part of a mine for the purposes of that Act; or

(b) a factory within the meaning of the Factories Act 1961(b).

This excludes mines and factories from the scope of these Regulations but the Regulations will be used as a form of code of practice by the Health and Safety Executive, Factory Inspectorate in respect of supplies and use of gas of a 'domestic' type within factories and mines (eg. office accommodation).

PART B: GAS FITTINGS - GENERAL PROVISIONS

Qualification and supervision

3. (1) No person shall carry out any work in relation to a gas fitting unless he is competent to do so.

> *'Competent' has not been defined but the person carrying out the work is responsible for ensuring that the Regulations are complied with.*

(2) The employer of any person carrying out such work shall ensure that paragraph (1) above is complied with.

(3) Without prejudice to the generality of paragraphs (1) and (2) above, no employer shall allow any of his employees to carry out any work in relation to a gas fitting, and no self-employed person shall carry out any such work, unless the employer or self-employed person, as the case may be, is a member of a class of persons approved for the time being by the Health and Safety Executive for the purposes of this paragraph.

(4) An approval given pursuant to paragraph (3) above (and any withdrawal of such approval) shall be in writing and notice of it shall be given to such persons in such manner as the Health and Safety Executive considers appropriate.

(5) The employer of any person carrying out any work in relation to a gas fitting shall ensure that the following provisions of these Regulations are complied with by that person.

> *Prohibits an employer from allowing his employees to carry out work in relation to gas fittings unless he (the employer) is a member of a class of persons approved by the Health and Safety Executive. Also prohibits a self-employed person from carrying out such work unless he is a member of such a class.*

Materials and workmanship

4. (1) No person shall install a gas fitting unless every part of it is of good construction and sound material and of adequate strength and size to secure safety.

(2) Without prejudice to the generality of paragraph (1) above, no person shall install in a building any pipe for use in the supply of gas which is -

(a) made of lead or lead alloy; or

(b) made of a non-metallic substance unless it is -

(a) 1954 c.70.
(b) 1961 c.34.

(i) a pipe connected to a readily movable gas appliance designed for use without a flue; or

Readily movable gas appliances should be taken to mean bunsen burners, lighting torches etc.; flexible pipes connecting such fittings need not be metallic.

(ii) a pipe entering the building and that part of it within the building is placed inside a metallic sheath which is so constructed and installed as to prevent, so far as is reasonably practicable, the escape of gas into the building if the pipe should fail.

(3) No person shall carry out any work in relation to a gas fitting otherwise than in a proper and workmanlike manner.

General safety precautions

5. (1) No person shall carry out any work in relation to a gas fitting in such a manner that gas could escape unless steps are taken to prevent any escape of gas which constitutes a danger to any person or property.

    (2) No person carrying out work in relation to a gas fitting shall leave the fitting unattended unless every incomplete gasway has been sealed with the appropriate fitting so as to be gastight or the gas fitting is otherwise safe.

    'Unattended' should be interpreted as meaning not present at the premises; however if the pipework system is not connected to a gas supply, it may be left unsealed as it is inherently safe by not being connected to a supply.

    If connected to a gas supply, the duty to seal upon leaving, even if only for a few minutes, is absolute.

    (3) Any person who disconnects a gas fitting shall, with the appropriate fitting, seal off every outlet of every pipe to which it was connected.

    If a 'gas fitting' – (which includes an appliance and a gas meter etc., Regulations 2(2) – is disconnected the gas supply must be capped or plugged.

    Should part of the installation pipework be disconnected then the open ends of pipework should also be sealed (capped or plugged).

    (4) No person carrying out work in relation to a gas fitting which involves exposing gasways which contain or have contained flammable gas shall smoke or use any source of ignition unless those gasways have been purged so as to remove all such gas or have otherwise been made safe from risk of fire or explosion.

(5) No person searching for an escape of gas from a gas fitting shall use any source of ignition.

(6) Where a person carries out any work in relation to a gas fitting which might affect the gas tightness of the gas supply system, he shall immediately thereafter test the system for gas tightness at least as far as the nearest valves upstream and downstream in the system.

In a domestic situation any 'work' on the installation pipework will mean that the whole installation should be tested for soundness. In a non-domestic situation the availability of upstream and downstream valves means that the amount of pipework to be tested can be reduced.

Protection against damage

6. (1) Any person installing a gas fitting shall ensure that every part of it is properly supported and so placed or protected as to avoid any undue risk of damage to the fitting.

(2) No person shall install a gas fitting if he has reason to suspect that foreign matter may block or otherwise interfere with the safe operation of the fitting, unless he has fitted to the gas inlet of, and any airway in the fitting a suitable filter or other suitable protection.

(3) No person shall install a gas fitting in a position where it is likely to be exposed to any substance which may corrode gas fittings unless the fitting is constructed of materials which are inherently resistant to being so corroded or it is suitably protected against being so corroded.

There are many potential sources of corrosion eg. solids, gases (atmospheres) and liquids.

Existing gas fittings

7. (1) No person shall make any alteration to any premises which would affect a gas fitting in such a manner that, if the fitting had been installed after the alteration there would have been a contravention or failure to comply with any provision of these Regulations in force.

Builders etc must now ensure that work carried out does not adversely affect the gas fittings already installed.

(2) No person shall do anything which would affect a gas fitting or any flue or means of ventilation used in connection with the fitting in such a manner that the subsequent use of the fitting might constitute a danger to any person or property.

Persons installing double glazing, heat recovery devices, cavity wall insulation etc., must now be aware of the effects that this would have on gas appliances and installations. Purpose provided ventilation may need to be added or examination put in hand to ensure that the alterations have not affected existing ventilation provisions.

Emergency controls

8. (1) No person shall give a new supply of gas for use in any building unless there is provided an emergency control to which there is adequate access situated -

    (a) if there is a dwelling to be supplied with gas in the building -

    (i) as near as is reasonably practicable to the point where the pipe supplying the gas enters the dwelling, and also

    (ii) if the pipe supplying the gas enters the building at a place not comprised within a dwelling, as near as is reasonably practicable to the point of entry, or

    (b) if there is no such dwelling in the building, as near as is reasonably practicable to the point where the pipe supplying the gas enters the building.

> *A 'new supply' to a building should be taken to mean a gas service to a building which had not previously had a supply of gas.*
>
> *This Regulation introduces the concept of the 'emergency control' – and requires the emergency control to be 'as near as is reasonably practicable to the point where the pipe supplying the gas enters the building (or dwelling)'.*
>
> *This means that provided there is adequate access to the emergency control it can be situated immediately inside or outside the building or dwelling. In most cases therefore the emergency control will take the form of a valve adjacent to and on the inlet side of the meter. ie. the meter control valve (See installation diagrams 1 and 2)*
>
> *In the flat or multiple occupancy type situation where the meter may be remote from the dwelling, eg. in a purpose designed common meter room, it will be necessary to install an emergency control at each individual dwelling. (See installation diagrams 4 and 5)*
>
> *The majority of emergency controls will be provided and labelled by the gas 'supplier'. There will be a responsibility placed on installers concerning the installation and labelling of emergency controls in certain limited applications, i.e. in flat or multiple occupancy situations such as those referred to above.*
>
> *Where the British Gas Region is the 'gas supplier' the necessary labels will be made available on request and a charge may be levied.*

(2) Any person installing an emergency control shall ensure that -

(a) any key, lever or hand wheel of the control is securely attached to the operating spindle of the control;

(b) any such key or lever is attached so that -

(i) the key or lever is parallel to the axis of the pipe in which the control is installed when the control is in the open position, and

(ii) where the key or lever is not attached so as to move only horizontally, gas cannot pass beyond the control when the key or lever has been moved as far as possible downwards; and

(c) either the means of operating the control are clearly and permanently marked or a notice in permanent form is prominently displayed near such means so as to indicate when the control is open and when the control is shut.

Suitable valves include those indicated in BS 1552

The 'emergency control' must have a notice explaining which position is open and which closed. The notice must be on or near the control.

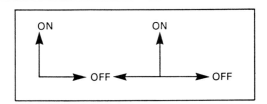

(3) Where a person installs an emergency control which is not to form part of a primary meter installation, he shall immediately thereafter prominently display on or near the means of operating the control a notice in permanent form bearing the words 'Gas Emergency Control' -

(a) indicating that the consumer should -

(i) shut off the supply of gas immediately in the event of an escape of gas in the building or dwelling, as the case may be, for which the control is provided;

(ii) where any gas continues to escape after the emergency control has been closed, as soon as practicable give notice of the escape to the supplier, and

(iii) not re-open the emergency control unit until all necessary steps have been taken to prevent gas from escaping again, and

(b) stating -

(i) the name of the supplier;

(ii) the emergency telephone number of the supplier, and

(iii) the date on which the notice was first displayed.

Where an emergency control is required downstream of the primary meter installation, it will be the installer's responsibility to provide such a control, and to ensure that all notices relating to emergency controls required by the Regulations are suitably positioned. (See installation diagrams 3, 4 and 5)

INSTALLATION DIAGRAMS

Fig. 1 External Meter Installation (eg. Meter box type)

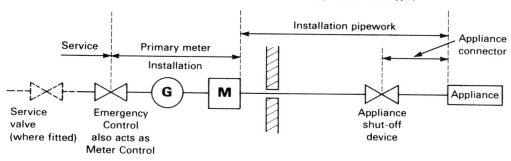

Labelling: Regulation 8(2)(c) and 14(1)

Fig. 2 Internal Meter Installation

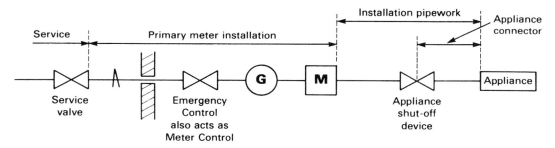

Labelling: Regulations 8(2)c, 8(3), 14(1)

A = Alternate position of Emergency Control

Fig. 3 Meter Remote from Dwelling (Meter Houses) or Detached Garage

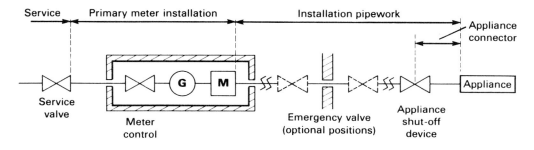

Labelling: Regulations 8(2)(c), 8(3), 14(1)

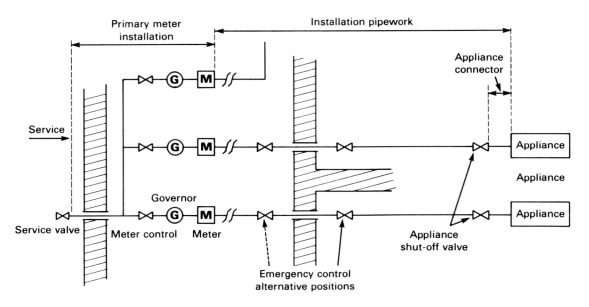

Fig. 4 Multi Occupancy Installation (Meters Remote from Dwelling)

Labelling: Regulations 8(2)c, 8(3), 14(1), 15(2)

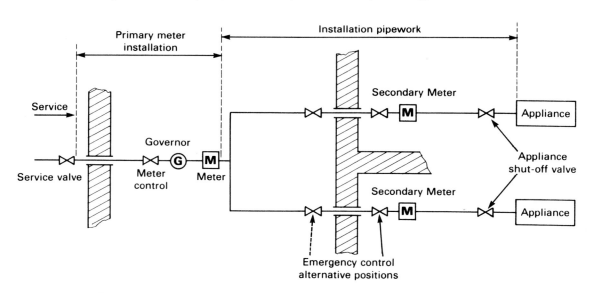

Fig. 5 Multi Occupancy Installation (with Secondary Metering)

Labelling: Regulations 8(2)c, 8(3), 14(1), 16

The notice should be completed at the time of installation of the emergency control, should show at least the month and year and should be displayed on or near the emergency control.

GAS EMERGENCY CONTROL

IN THE EVENT OF AN ESCAPE

Turn off supply at emergency control valve. Open windows.

Do **NOT** search with naked light and if gas escape persists

Do **NOT** turn on gas until the Escape has been repaired.

THE EMERGENCY TELEPHONE NUMBER IS:

Date _____

Electrical continuity - general

9. In any case where it is necessary to avoid danger, no person shall carry out work in relation to a gas fitting without first providing a suitable bond to maintain electrical continuity until the work is completed.

'Suitable bond' means an electrical connection made to bridge a gap caused by the temporary absence of a continuous single gas pipe or gas fitting. It should be used when disconnecting and reconnecting pipework in those instances where production of a spark could cause a hazard; this action will safeguard against the risk of fire, explosion or electric shock caused by contact with other services.

PART C: METER INSTALLATIONS

Interpretation of Part C

10. In this Part -

    'meter box' means a receptacle or compartment designed and constructed to contain a meter with its associated gas fittings;

    'secondary meter' means a meter for ascertaining the quantity of gas provided by a person supplied through a primary meter for use by another person.

The definition of 'secondary meter' excludes customers own check meters (eg. those installed for fuel efficiency and cost monitoring purposes).

Meters - general provisions

11. (1) No person shall install a meter on or under a stairway or in any other part of a building with two or more floors above the ground floor, where the stairway or that part of the building provides the only means of escape in case of fire, unless the meter replaces an existing meter and sub-paragraph (a) or (b) of paragraph (2) below is complied with.

The action to be taken in respect of a meter installation for building with two or more floors above ground floor (3 storey buildings and above) is summarised below.

__New Installations:__ No meter may be installed on or under stairs or elsewhere if the stairway or other part of the building provides the sole means of escape in the event of fire.

__Replacement Installation:__ If an existing meter is fitted on or under a stairway or elsewhere either of which is on the sole escape route, one of the following must be carried out if the meter is to be changed.

either

(a) *Install a fire resistant meter*

   OR

(b) *install a meter housed in a fire resistant compartment which has doors fitted with automatic self-closing devices*

   OR

(c) *Fit a thermal cut-off valve upstream of the meter and governor to operate at 95°C ambient temperature*

   OR

(d) *Resite the meter in compliance with new installations guidance.*

(2) No person shall install a meter in a building with no more than one floor above the ground floor on or under a stairway or in any other part of the building, where the stairway or that other part of the building provides the only means of escape in case of fire, unless -

(a) the meter is -

   (i) of fire resisting construction; or

   (ii) housed in a compartment with automatic self-closing doors and which is of fire resisting construction; or

(b) the pipe immediately upstream of the meter or, where a governor is adjacent to the meter, immediately upstream of that governor, incorporates a device designed to cut off automatically the flow of gas if the temperature of the device exceeds 95°C.

The following summarises the action for buildings with less than two floors above the ground floor (2 storey building and below)

New Installations: *If there is no alternative to fitting the meter on or under the stairs or on the route of the sole means of escape in case of fire, carry out one of the courses of action set out in 11 'Replacement Installation' (a)(b)(c) above*

Replacement Installations: *Replace in accordance with instructions under 'Replacement Installation' in 11 above*

(3) In paragraph (2)(a) above, the expression 'fire resisting construction' means that, if the meter or the compartment housing the meter were subjected for 30 minutes to the furnace test described in British Standard BS 476 (Fire Tests on Building Materials and Structures) Part 8: 1972 (Test methods and criteria for the fire resistance of elements of building construction) ISBN: 0 580 07166 9 as amended by Amendment Slip No. 1 published 30th January 1976 and Amendment No. 2 published and effective 30th November 1981 or an equivalent test, the construction of the meter would not be so adversely affected that gas could escape in hazardous quantities.

(4) No person shall install a meter unless the installation is so placed as to ensure that there is no risk of damage to it from electrical apparatus.

BS 6400 - Installation of Domestic Gas Meters - meet the fire resisting requirements (fire resisting partition made of an electrically insulated material placed between gas and electric meters and their controls if they are within 150mm (6ins) of each other).

(5) No person shall install a meter except in a readily accessible position for inspection and maintenance.

(6) Where a meter has bosses or side pipes attached to the meter by a soldered joint only, no person shall make rigid pipe connections to the meter.

On 'Tin Case' meters at least one connection must be semi-rigid (eg an 'Anaconda' type connection).

(7) Where a person installs a meter and the pipes and other gas fittings associated with it, he shall ensure that -

(a) immediately thereafter they are adequately tested to verify that they are gas tight and examined to verify that they have been installed in accordance with any provisions of these Regulations in force; and

(b) immediately after such testing and examination, purging is carried out throughout the meter and every other gas fitting through which gas can then flow so as to remove safely all air and gas other than the gas to be supplied.

After a meter has been installed it must be visually checked to ensure that its installation complies with these Regulations; it must be tested for gas tightness and purged through the whole installation, including any commissioned appliances. 'Gas tightness' means to Gas Industry Standards.

Meter boxes

12. (1) Where a meter is housed in a meter box attached to or built into the external face of the outside wall of a building, the meter box shall be so constructed and installed that any gas escaping within the box cannot enter the building or any cavity in the wall but must disperse to the external air.

    Particular care must be taken to adequately sleeve and seal the installation pipe in situations where it passes out of the rear of the meter box and enters the building via the cavity.

    (2) No person shall knowingly store flammable materials in any meter box.

    Meter rooms, compartments and boxes should not be used as store rooms for flammable materials (eg. paints, rags, petroleum spirits, LPG etc.)

    (3) No person shall install a meter in a meter box provided with a lock, unless the consumer has been provided with a key to the lock clearly labelled 'Gas Meter Box' in black capital letters on a yellow ground.

    The 'gas supplier' or meter box manufacturer should supply a suitably marked key. The duty remains with the installer of the meter to ensure that the consumer has been supplied with this suitably marked key.

Governors

13. (1) No person shall install a primary meter or a meter bypass used in connection with a primary meter unless -

    (a) there is a governor regulating the pressure of gas supplied through the meter or the bypass, as the case may be, which provides adequate automatic means for preventing the gas fittings connected to the downstream side of the governor from being subjected to a pressure greater than that for which they were designed;

    All supplies must be governed by either a meter governor or service governor but not necessarily both. On low pressure installations the intent of this Regulation will be met by the use of normal low pressure governor having 'lock up' facility.

    (b) where the normal pressure of the gas supply is 75 millibars or more there are also adequate automatic means for preventing, in case the governor should fail, those gas fittings from being subjected to such a greater pressure; and

(c) where the governor contains a relief valve or liquid seal, such valve or seal is connected to a vent pipe of adequate size so installed that it is capable of venting safely.

(2) Where a person installs a governor for regulating the pressure of gas through a primary meter or a meter bypass used in connection with a primary meter, he shall immediately thereafter adequately seal the governor to prevent its setting from being interfered with without breaking of the seal.

(3) No person except the supplier of the gas or a person authorised to act on his behalf shall break a seal applied under paragraph (2) above.

Installers are not permitted to break the seal on a meter governor unless authorised (in writing) to do so by British Gas where it is the 'supplier'. Similar permission would have to be obtained from other 'gas suppliers' to break the seals of their governors where they supply the gas.

The governor must be resealed after any work (eg resetting the pressure). Cases where the meter governors are not giving the correct working pressure must be reported to the 'gas supplier' before any action.

Meters - emergency notices

14. (1) No person shall supply gas through a primary meter installed after the commencement of this Regulation unless he ensures that a notice in permanent form is prominently displayed on or near the meter -

    (a) indicating that the consumer should -
        (i) shut off the supply of gas immediately in the event of an escape of gas in the consumer's premises;
        (ii) where any gas continues to escape after the supply has been shut off, immediately give notice of the escape to the supplier; and
        (iii) not re-open the supply until all necessary steps have been taken to prevent the gas from escaping again, and
    (b) stating -
        (i) the name of the supplier,
        (ii) the emergency service telephone number of the supplier, and
        (iii) the date on which the notice was first displayed.

This is the responsibility of the gas supplier. The meter label must be completed at the time of the installation of the meter and show at least the month and year.

(2) Where a meter is installed in a building at a distance of more than 2 metres from, or out of sight of, the nearest upstream emergency control in the building, no person shall supply or provide gas through the meter unless he ensures that a notice in permanent form is prominently displayed on or near the meter indicating the position of that control.

This new Regulation requires a permanent notice at the meter when the Emergency Control is more than 2 metres from, or out of site of the meter. The notice should include sufficient information to enable any person who goes to the meter to find the location of the emergency control.

Primary meters

15. (1) No person shall install a prepayment meter as a primary meter through which gas passes to a secondary meter.

(2) Any person who first supplies gas through any service pipe after the commencement of this Regulation to more than one primary meter shall ensure that a notice in permanent form is prominently displayed on or near each primary meter indicating that this is the case.

Typical primary meter label

```
┌─────────────────────┐
│        GAS          │
│   Joint Service to  │
│   Primary Meters    │
└─────────────────────┘
```

(3) Where a primary meter is removed, the person who last supplied gas through the meter before removal shall -

(a) where the meter is not forthwith re-installed or replaced by another meter -

   (i) close any service valve which controlled the supply of gas to that meter and did not control the supply of gas to any other primary meter; and

   (ii) clearly mark any live gas pipe in the premises in which the meter was installed to the effect that the pipe contains gas; and

(b) where the meter has not been re-installed or replaced by another meter before the expiry of the period of 12 months beginning with the date of removal of the meter and there is no such service valve as is mentioned in sub-paragraph (a)(i) above, ensure that the service pipe for those premises is disconnected as near as is reasonably practicable to the main and that any part of the pipe which is not removed is sealed at both ends with the appropriate fitting.

In (i) above, the house entry tee and meter control in an external box are considered to be service valves.

In (a) above, when a primary meter is removed and not immediately replaced the service valve should be closed and the service inside the premises marked with self adhesive yellow tape with black letters '...GAS...'

Secondary meters

16. Any person providing gas through a secondary meter shall ensure that a notice in permanent form is prominently displayed on or near the primary meter indicating the number and location of secondary meters installed.

Landlords using secondary meters are responsible for labelling of the primary meter. Installers should bring this Regulation to the landlord's attention. The notice should indicate the number of secondary meters with location details. (See installation diagram 5)

PART D: INSTALLATION PIPES

Safe use of pipes

17. (1) No person shall install an installation pipe in any position in which it cannot be used with safety having regard to the position of other pipes, drains, sewers, cables, conduits and electrical apparatus and to any parts of the structure of any building in which it is installed which might affect its safe use.

(2) Any person who connects an installation pipe to a primary meter installation shall, in any case where electrical cross-bonding may be necessary, inform the responsible person that such cross-bonding should be carried out by a competent person.

*The installer's responsibility is to **inform** the responsible person e.g. owner, occupier or manager that electrical cross-bonding is or may be needed and must be carried out by a competent person, e.g. an approved contractor of the National Inspection Council for Electrical Installation Contractors (NICEIC) or the local Electricity Board.*

Suitable wording for a standard letter has been agreed and installers are recommended to use this or similar wording when notifying the responsible person:-

> "Some types of electrical installations are fitted with cross-bonding, which is the connection of the internal gas and water pipes to the installation's earth terminal. In particular those installations with P.M.E. (protective multiple earth) must, by law, be fitted with cross-bonding.
>
> The new gas installation pipe now fitted in your premises does not appear to be cross-bonded to the electrical installation.
>
> I am required by the Gas Safety (Installation and Use) Regulations 1984 to tell you that any necessary cross-bonding should be carried out by a competent person.
>
> I advise that you have this checked by a NICEIC Approved Contractor or by your Electricity Board.
>
> If you are a tenant of this property, would you please bring this matter to the attention of the owner or his agent."

Enclosed pipes

18. (1) No person shall install any part of an installation pipe in a wall or a floor or standing of solid construction unless it is so constructed and installed as to be protected against failure caused by movement.

Accepted methods of laying pipework in solid structures should satisfy this Regulation:-

- *pipes laid on concrete foundations and covered with a screed*
- *pipes placed in preformed channels*
- *pipes wrapped with a resilient soft material*
- *soft copper inserted through plastic tube set in the floor slabs or wall*

(2) No person shall install an installation pipe so as to pass through a wall or a floor or standing of solid construction from one side to the other unless any part of the pipe within such wall, floor or standing as the case may be -

(a) takes the shortest practicable route; and

(b) is enclosed in a gastight sleeve and the pipe and the sleeve are so constructed and installed as to prevent, as far as is reasonably practicable having regard to paragraph (1) above, gas passing along any space between the pipe and the sleeve or between the sleeve and such a wall, floor or standing as the case may be.

This Regulation is designed to protect installation pipes against failure due to movement.

(3) No person shall install any part of an installation pipe in the cavity of a cavity wall unless the pipe is to pass through the wall from one side to the other.

These Regulations should not be read in isolation and in particular 18(3) must be considered with 18(2) ie. a pipe passing through a cavity wall must be sleeved.

(4) No person shall install an installation pipe under the foundations of a building or in the ground under the base of a wall or footings.

This Regulation should be taken as refering to load bearing walls only.

(5) No person shall install an installation pipe in an unventilated shaft, duct or void.

Protection of buildings

19. No person shall install an installation pipe in a way which would impair the structure of a building or impair the fire resistance of any part of its structure.

    An example of bad practice for this Regulation would be if a pipe passed through a fire wall without the pipe and sleeve being fire stopped and sealed in position.

Clogging precautions

20. No person shall install an installation pipe in which deposition of liquid or solid matter is likely to occur unless a suitable vessel for the reception of any deposit which may form is fixed to the pipe in a conspicuous and readily accessible position and safe means are provided for the removal of the deposit.

    Generally, under normal circumstances, condensate traps and filters are not necessary.

Testing and purging of pipes

21. (1) Where a person carries out work in relation to an installation pipe which might affect the gastightness of any part of it, he shall immediately thereafter ensure that -

    (a) that part is adequately tested to verify that it is gastight and examined to verify that it has been installed in accordance with those provisions of these Regulations in force; and

    (b) after such testing and examination, any necessary protective coating is applied to the joints of that part.

'Gastight' above means to Gas Industry Standards. The pipe and fitting must be tested to those standards.

The purpose in (b) above is to remove any possibility of an erroneous result due to the protective coating (such as paint) temporarily sealing a leak and the requirement applies to the particular part of the pipe or fitting rather than the whole pipe.

(2) Where gas is being supplied to any premises in which an installation pipe is installed and a person carries out work in relation to the pipe, he shall also ensure that -

(a) immediately after complying with the provisions of sub-paragraphs (a) and (b) of paragraph (1) above, purging is carried out throughout every installation pipe through which gas can then flow so as to remove safely all air and gas other than the gas to be supplied;

(b) immediately after such purging, if the pipe is not to be put into immediate use, it is sealed off at every outlet with the appropriate fitting;

(c) if such purging has been carried out through a loosened connection, the connection is retested for gastightness after it has been retightened; and

(d) every seal fitted after such purging is tested for gastightness.

These requirements apply at all times to installation pipes in which gas is supplied, not just when gas is first supplied.

(3) Where gas is not being supplied to any premises in which an installation pipe is installed at a time when a person carries out work in relation to the pipe, no person shall supply gas to the premises unless he has caused such purging and other work as is specified in sub-paragraphs (a) to (d) of paragraph (2) above to be carried out.

These requirements apply to installation pipes which are not supplied with gas at the time of installation or when work is being carried out on them.

Marking of pipes

22. (1) Any person installing, elsewhere than in any premises or part of premises used only as a dwelling or for living accommodation, a part of an installation pipe which is accessible to inspection shall permanently mark that part in such a manner that it is readily recognisable as part of a pipe for conveying gas.

Pipes installed in non-domestic premises must be colour identified (BS 1710). The colour for Natural Gas is yellow ochre. This is a fairly flexible Regulation, but whatever the means, the pipe must be readily identifiable

(2) The responsible person for the premises in which any such part is situated shall ensure that the part continues to be so recognisable so long as it is used for conveying gas.

Large consumers

23. (1) Where the service pipe to any building having two or more floors to which gas is supplied or (whether or not it has more than one floor) a floor having areas with a separate supply of gas, has an internal diameter of 50mm or more, no person shall install an incoming installation pipe to any of those floors or areas as the case may be unless -

   (a) a valve is installed in the pipe in a conspicuous and readily accessible position; and

   (b) a line diagram in permanent form is attached to the building in a readily accessible position as near as practicable to the primary meter indicating the position of all installation pipes of internal diameter of 25mm or more, meters, emergency controls, valves, pressure test points, condensate receivers and electrical bonding of the gas supply systems in the building.

   (2) In paragraph (1)(b) above 'pressure test point' means a gas fitting to which a pressure gauge can be connected.

This is an attempt to get two conditions into one Regulation. It may be clearer to read it as two separate clauses:-

Where the service pipe to any building with two or more floors to which gas is supplied - has an internal diameter of 50mm or more - no person shall install an incoming installation pipe to any of those floors unless:- (a) and (b) are complied with.

or

Where the service pipe to any building is 50mm or more and the building has floor areas with a separate supply of gas, no person shall install an incoming installation pipe to any of those areas unless (a) and (b) are complied with.

This Regulation ensures that the main sections of the gas installation, its emergency controls and isolating valves etc., can be identified and located in the event of an emergency. It is unnecessary to show pipes below 25mm (1'') diameter.

It should be noted that the Regulation refers to the number of floors and service pipe diameter, not the size of the meter.

It is the responsibility of the installer to provide the permanent durable diagram when first installing supplies to a separate floor or area.

The Regulation is not retrospective. However, if a supply is provided to a new area or floor, a diagram is required for the whole installation, or, any existing diagram should be updated. Extensions within any floor or area do not require a new diagram but consideration should be given to updating any existing diagram.

Gas Industry recommendations on the format of the diagram are given below.

1. Size — For the simplest installations A4 paper size. Where they are more complicated use not less than A3.

2. Materials — Ordinary drawing materials may be used but some form of protection should be provided, for example, glass or plastic.

3. Location — One copy of the diagram must be placed near the primary meter; consideration should be given to having copies elsewhere, for example, at the gatehouse.

4. Updating — Whenever changes are made to the pipework installation the line diagram **must** be updated. It is suggested that the updating of the diagram is made the responsibility of the Head of Department which handles services, for example, the Maintenance Engineer or Services Engineer. It should be noted that the line diagram is required to enable rapid location of valves for emergency purposes; an out-of-date diagram could therefore be a serious hazard.

5. Detail — The diagram, preferably based on a site or floorplan, should show sufficient detail to enable location of manual valves in an emergency. It is therefore not necessary to show every last connection. For example if a building had 76mm or 102mm pipes to appliances with 19mm tees for supplying a hot water boiler or canteen, only the valved tee would be shown, not the detailed pipe run; in contrast if the only equipment in a building was a boiler supplied by 102mm lines from a 102mm service, the full layout would have to be shown. The figure shows an example of a line diagram; note that a key should always be included.

This diagram is an illustration only and does not purport to depict good pipe layout.

It is recommended that consideration is given to following these line diagram requirements in factory installations etc.

CITB Co. LTD. GAS SUPPLY LINE DIAGRAM

This diagram has been prepared to conform with the Gas Safety (Installation and Use) Regulations 1984 and must be displayed adjacent to the primary gas meter.

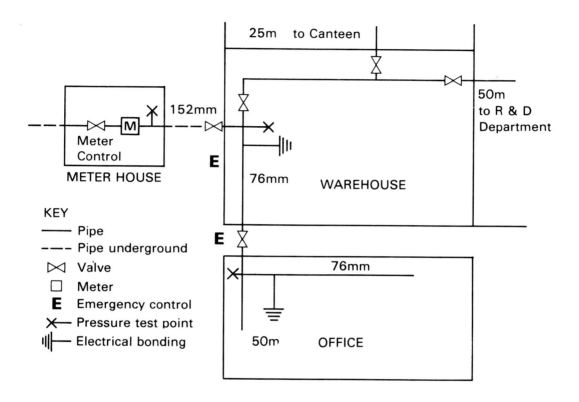

NOTE: Copies of this diagram are displayed as follows:
1. Meter House
2. Gate House
3. Engineers

Note: Pipes below 25mm not shown

PART E: GAS APPLIANCES

*PART E of the Regulations clearly recognises the importance of always checking appliances in respect of flueing, ventilation and combustion. Regulation 25(1) refers to the checks being made when **INSTALLING** the appliance, Regulation 25(9) refers to the checks made when **SERVICING** the appliance. Regulation 33(1), (2) and (3) refers to **TESTING AND EXAMINING** the appliance and Regulation 34(1) places responsibilities on persons who know or have reason to suspect that flueing, ventilation and/or combustion are not effective when the appliance is **BEING USED**.*

Interpretation of Part E

24.   In this Part -

'flue pipe' means pipe forming a flue but does not include a pipe built as a lining into either a chimney or a gas appliance ventilation duct;

'heat input', in relation to a gas appliance, means the gas consumption of the appliance expressed in terms of the quantity of heat supplied to the appliance in a specific time;

'operating pressure', in relation to a gas appliance, means the pressure of gas at which it is designed to operate.

Gas appliances - safety precautions

25.   (1) No person shall install a gas appliance unless -

   (a)   the appliance and the gas fittings and other works for the supply of gas to be used in connection with the appliance,

   (b)   the means of removal of the products of combustion from the appliance,

   (c)   the availability of sufficient permanent supply of air for the appliance for proper combustion,

   (d)   the means of ventilation to the room or internal space in which the appliance is to be used, and

   (e)   the general conditions of installation including the stability of the appliance and its connection to any other gas fitting,

are such as to ensure that the appliance can be used without constituting a danger to any person or property.

Adherence to British Standards Codes of Practice referring to appliance installation should ensure compliance with this Regulation. (British Standards are obtainable from: B.S.I. Sales Department, Linford Wood, Milton Keynes, MK14 6LE). British Gas also provide technical information on specific appliances and installations which are publicised from time to time by CORGI and are available from British Gas Regional offices.

The requirements of Regulations 25(1) also apply to the installation of secondhand appliances.

(2) No person shall connect a flued domestic gas appliance to the gas supply system except by a permanently fixed rigid pipe.

This is to ensure that flued appliances remain connected to their flues.

(3) No person shall install a used gas appliance without verifying that it is in a safe condition for further use.

Installers of secondhand appliances must take special care to ensure that the appliance is safe to use before reconnection. The Regulation also applies to appliances when installed in another location.

(4) No person shall install a gas appliance which does not comply with any enactment imposing a prohibition or restriction on the supply of such an appliance on grounds of safety.

(5) No person carrying out the installation of a gas appliance shall leave it connected to the gas supply unless the appliance can be used safely.

If the appliance is unsafe after installation, it must be immediately disconnected. Gas appliances must not be connected to the live supply until such time as they can be fully commissioned.

(6) No person shall install a gas appliance without there being at the inlet to it means of shutting off the supply of gas to the appliance unless the provision of such means is not reasonably practicable.

This Regulation will be satisfied by the fitting of a valve at the inlet to each appliance. The valve need not be operable by the appliance 'user'.

(7) No person shall carry out any work in relation to a gas appliance which bears an indication that it conforms to a type approved by any person as complying with safety standards in such a manner that the appliance ceases to comply with those standards.

The purpose of this Regulation is to prevent unauthorised modifications of appliances which may reduce the safety standards of those appliances; it does not preclude safety modifications which do not reduce the standards of the appliance.

(8) No person carrying out work in relation to a gas appliance which bears an indication that it so conforms shall remove or deface the indication.

(9) Where a person services a domestic gas appliance, he shall immediately thereafter examine:-

(a) the effectiveness of any flue,
(b) the supply of combustion air,
(c) its heat input and operating pressure, and
(d) its safe functioning,

and forthwith notify the responsible person for the premises in which the appliance is situated of any defect.

It is essential that (a) to (d) is undertaken immediately following the service. Although the word 'domestic' is used in the Regulation it is recommended that this practice is applied to all appliances.

Thus as a part of any service work schedule the listed examination must be carried out and any defects must be notified to the 'Responsible Person' (e.g. the occupant, landlord or tenant) without delay. In this context 'domestic gas appliances' should be taken to mean those appliances with an input rating not exceeding 60kW intended for use in domestic premises. For servicing non-domestic appliances Regulation 34 is more applicable, and service schedules should reflect this.

Ensuring that the appliance is set at the manufacturers' correct operating pressure will satisfy Regulation 25(9)(c) unless the flame picture or performance of the appliance appears suspect – in which case a check must be made.

Although not a specific requirement of these Regulations, it is recommended that Installers aware of any potentially dangerous installation should, in addition to advising the 'responsible person' not to use the equipment, confirm this in writing, indicating the work necessary to make the installation safe for use; and should inform the Regional office of the Health and Safety Executive.

Building legislation

26. No person shall install a gas appliance unless the appliance and the gas fittings and any flue or means of ventilation to be used in connection with the appliance comply with -

   (a) in the case of an installation in Greater London other than an outer London borough,

   (i) any provision of the London Building Acts 1930 to 1939 (**a**) and any byelaws made thereunder, and

   (ii) any provision of the London Gas Undertakings (Regulations) Act 1939 (**b**) and any regulations made thereunder;

(b) in the case of an installation in any part of England or Wales, any provision of regulations made or having effect under section 61(1) (power to make building regulations) of the Public Health Act 1936 (**c**); or

(c) in the case of an installation in Scotland, any provision of regulations made under section 3(1) (building standards regulations) of the Building (Scotland) Act 1959 (**d**),

which is in force at the date of installation.

Flues

27. (1) No person shall install a gas appliance to any flue unless the flue is suitably constructed and in a proper condition for the safe operation of the appliance.

(2) No person shall install a flue pipe so that it enters a brick or masonry chimney in such a way that the seal between the flue pipe and chimney cannot be inspected.

(3) No person shall connect a gas appliance to a flue which is surrounded by an enclosure unless that enclosure is so sealed that any spillage of products of combustion cannot pass from the enclosure to any room or internal space other than the room or internal space in which the appliance is installed.

The objective of this Regulation is to prevent the spillage of the products of combustion via the enclosure, to any room or internal space other than the one in which the appliance is situated.

(4) No person shall install a power operated flue system for a gas appliance unless it safely prevents the operation of the appliance if the draught fails.

Access

28. No person shall install a gas appliance except in such a manner that it is readily accessible for operation, inspection and maintenance.

Manufacturer's instructions

29. Any person who installs a gas appliance shall leave with the owner or occupier of the premises in which the appliance is installed all instructions provided by the manufacturer accompanying the appliance.

Room-sealed appliances

30. No person shall install a gas appliance in a private garage or in a bath or shower room unless it is a room-sealed appliance.

Building Regulation requirement.

Suspended appliances

31. No person shall install a suspended gas appliance unless the installation pipe to which it is connected is so constructed and installed as to be capable of safely supporting the weight imposed on it and the appliance is designed to be so supported.

The appliance manufacturers installation instructions will specify the method of supporting a particular appliance and whether it is acceptable to mount the appliance on the correctly supported pipework.

Flue dampers

32. (1) Any person who installs an automatic damper to serve a gas appliance shall -

   (a) ensure that the damper is so interlocked with the gas supply to the burner that burner operation is prevented in the event of failure of the damper when not in the open position, and

   (b) immediately after installation examine the appliance and the damper to verify that they can be used together safely without constituting a danger to any person or property.

See British Gas publication IM/19. Automatic Flue Dampers for use with Gas-Fired Space Heating and Water Heating Appliances.

(2) No person shall install a manually operated damper to serve a domestic gas appliance.

(3) No person shall install a domestic gas appliance to a flue which incorporates a manually operated damper unless the damper is permanently fixed in the open position.

In particular this Regulation applies where a gas fire is fitted to a solid fuel chimney. The damper must either be removed or permanently fixed in the open position.

Testing of appliances

33. (1) Where a person installs a gas appliance at a time when gas is being supplied to the premises in which the appliance is installed, he shall immediately thereafter test its connection to the installation pipe to verify that it is gastight and examine the appliance and the gas fittings and other works for the supply of gas and any flue or means of ventilation to be used in connection with the appliance for the purpose of ascertaining whether -

   (a) the appliance has been installed in accordance with these Regulations;

(b) the heat input and operating pressure are as recommended by the manufacturer;

(c) the appliance has been installed with due regard to any manufacturer's instructions provided to accompany the appliance; and

(d) all gas safety controls are in proper working order.

With 'due regard' in (c) above means to take account of a situation where the manufacturers instructions may not be practicable, but where equally safe methods could be employed.

Regulations 33(1) (a) to (d) - apply at all times when gas is being supplied - not only when it is first supplied.

(2) Where a person carries out such testing and examination in relation to a gas appliance and adjustments are necessary to ensure compliance with the requirements specified in sub-paragraphs (a) to (d) of paragraph (1) above, he shall either carry out those adjustments or disconnect the appliance from the gas supply.

'Such testing and examination' refers to 33(1) (a) to (d). This clause is of major importance for those commissioning equipment and installations. The Regulations make adequate, incomplete or incorrect commissioning of gas combustion equipment or gas safety controls an offence.

The following are among the appropriate checks for domestic appliances but the list should not be regarded as complete:-

- *Correct calorific value of the gas for which the appliance is intended. (Should be checked before installation)*
- *Correct governor operation and pressure setting*
- *Actual gas flow rate*
- *Visual check of the flame picture.*

(3) Where a person installs a gas appliance in any premises at a time when gas is not being supplied to the premises, no person shall supply gas to that appliance unless he has caused such testing and examination and adjustments as are specified in paragraphs (1) and (2) above to be carried out.

This Regulation takes account of the fact of the time delay which sometimes occurs between the installation of appliances etc. and the provision of a gas supply.

The supplier may merely make a supply of gas available, then notify the responsible person that a supply of gas is available and that the installer should now return to connect and commission the appliance(s).

It is recommended therefore that the installer makes provision for a connecting, commissioning and testing visit in any contract or arrangement with the customer. '...no person shall supply gas to that appliance...' means that gas can be supplied to other fully tested appliances in the premises.

Unsafe appliances

34. (1) No person shall use a gas appliance or permit a gas appliance to be used if at any time he knows or has reason to suspect -

    (a) that there is insufficient supply of air available for the appliance for proper combustion at the point of combustion;

    (b) that the removal of the products of combustion from the appliance is not being or cannot safely be carried out;

    (c) that the room or internal space in which the appliance is situated is not adequately ventilated for the purpose of providing air containing a sufficiency of oxygen for the persons present in the room, or in, or in the vicinity of, the internal space while the appliance is in use;

    (d) that any gas is escaping from the appliance or from any gas fitting used in connection with the appliance; or

    (e) that the appliance or any part of it or any gas fitting or other works for the supply of gas used in connection with the appliance is so faulty or maladjusted that it cannot be used without constituting a danger to any person or property.

    These Regulations apply to all appliances whenever installed.

    If a gas appliance or gas installation is in an unsafe condition, the responsibility for remedying the defect and making safe or stopping the use rests with the 'responsible person' (occupier or owner of the premises).

    It is recommended that if the installer is aware of any potential dangerous installation, in addition to advising the 'responsible person' not to use the equipment, he should confirm this in writing and indicate the work necessary to make the installation safe.

    (2) Any person engaged in carrying out any work in relation to a gas main, service pipe or gas fitting who knows or has reason to suspect that any defect or other circumstance referred to in paragraph (1) above exists shall forthwith take all reasonably practicable steps to inform the responsible person for the premises in which the appliance is situated or, where that is not reasonably practicable, the supplier of gas to the appliance.

    It is the intention of this Regulation that if the private installer cannot advise the 'responsible person' (occupier or owner of the premises) that the dangerous situation exists then the supplier of gas (as the supplier) should be advised.

(3) In paragraph (2) above the expression 'work' shall be construed as if, in the definition of 'work' in Regulation 2(1) above, every reference to a gas fitting were a reference to a gas main, service pipe or gas fitting.

This Regulation extends the definition of 'work' given in Regulation 2(1) to include work on gas mains and service pipes in addition to work on gas fittings and it thus extends the scope of the Regulation to those engaged in work on distribution and transmission.

PART F: MISCELLANEOUS

Escape of gas

35. (1) If the responsible person for any premises knows or has reason to suspect that gas is escaping into those premises, he shall immediately take all reasonable steps to cause the supply of gas to be shut off at such place as may be necessary to prevent further escape of gas.

Responsibility for taking action in the event of an escape lies with the 'responsible person'.

(2) If gas continues to escape into those premises after the supply of gas has been shut off or when a smell of gas persists, the responsible person for the premises discovering such escape or smell shall immediately give notice of the escape or smell to the supplier of gas to the premises.

Any uncontrollable escape of gas must be reported immediately to the gas supplier.

(3) Where an escape of gas has been stopped by shutting off the supply, no person shall cause or permit the supply to be re-opened until all necessary steps have been taken to prevent gas from escaping again.

This Regulation is not intended to prevent a controlled restoration of the gas supply to facilitate leak detection.

Penalty

36. Subject to Regulation 37 below, a person contravening or failing to comply with any provision of these Regulations in force shall be guilty of an offence and liable on summary conviction to a fine not exceeding £2,000.

In England and Wales prosecution can only be instituted by or with the consent of the Secretary of State or by the Director of Public Prosecution.

In Scotland prosecution can only be instituted by or with the consent of the Secretary of State or by the Procurator Fiscal.

Exception as to liability

37. No person shall be guilty of an offence by reason of any contravention of, or failure to comply with, Regulation 3(2), 4(1), 6(3), 14, 15(2) or (3), 16 or 33(1) in any case in which he can show that he took all reasonable steps to prevent the contravention or failure.

Amendment of Gas Safety Regulations 1972 and Gas Safety (Rights of Entry) Regulations 1983

38. (1) The Gas Safety Regulations 1972(**a**) shall be amended as follows:-

    (a) Parts III to VI shall cease to have effect.

    (b) In Regulation 49, for the words 'gas pipe, pipe fitting or meter' there shall be substituted, in both places where they appear, the words 'service pipe or associated pipe fitting' and the words from 'except' to the end shall be omitted.

    (c) In Regulation 50, for the words from 'gas fitting' to 'premises' there shall be substituted the words 'service pipe or associated pipe fitting'.

    (d) In Regulation 51 -

        (i) in paragraph (1), for the words from 'gas fitting' (where those words first appear) to 'premises' and for the words from 'gas fitting' (where those words secondly appear) to 'question' there shall be substituted the words 'service pipe or associated pipe fitting' and for the words 'Parts II to V' there shall be substituted the words 'Part II';

        (ii) in paragraph (2), for the words from 'gas fitting' to 'premises' there shall be substituted the words 'service pipe or associated pipe fitting' and the words 'III, IV or V (as the case may be)' and the proviso shall be omitted;

        (iii) in paragraph (3), for the words from 'gas fitting' to 'premises' there shall be substituted the words 'service pipe or associated pipe fitting'; and

        (iv) in paragraphs (4) and (5), for the words from 'gas fitting' to 'premises' there shall be substituted the words 'service pipe or associated pipe fitting' and for the words 'Part II to V' there shall be substituted the words 'Part II';

        (v) after paragraph (5), the following paragraph shall be added -

            '(6) A person who makes any such replacement of a service pipe shall ensure that, as soon as is reasonably practicable, any part of the old pipe which is not removed is disconnected as near to the main as is reasonably practicable.'

(2) Regulation 8 of the Gas Safety (Rights of Entry) Regulations 1983(**a**) shall be amended by substituting for the expression '£1,000' the expression '£2,000'.

The purpose of this Regulation is to amend certain sections of the Gas Safety Regulations 1972 which will remain in force until regulations covering transmission and distribution of gas are made.

The amendments to the Gas Safety Regulations 1972 to which this regulation refers are reproduced below:-

49. An electrical connection, shall be maintained by means of temporary continuity bonding while a service pipe or associated pipe fitting is being removed or replaced until the work is disconnecting or connecting the service pipe or associated pipe fitting, as the case may be, has been completed.

50. A person who disconnects a service pipe or associated pipe fitting shall seal it off, cap it or plug it at every outlet of every pipe to which it is connected with the appropriate pipe fitting.

51. (1) No alteration shall be made to a service pipe or associated pipe fitting (whether it has been installed before or after the date of coming into operation of these regulations) if as a result of such alteration there would have been a contravention of or failure to comply with any provision of Part II of these regulations if the service pipe or associated pipe fitting had been installed at the date of the alteration.

    (2) On every replacement of a service pipe or associated pipe fitting (whether it has been installed before or after the date of coming into operation of these regulations) the provisions of Part II, of these regulations shall apply to its replacement as they apply to its installation after the said date:

    (3) A person who makes any alteration to or replacement of a service pipe or associated pipe fitting shall ensure that it is forthwith after such testing examined to verify that there would have been no such contravention of or failure to comply with any provision of Parts II of these regulations as is referred to in paragraph (1).

    (5) A person who makes any such replacement of a service pipe or associated pipe fitting shall ensure that it is forthwith after such testing examined to verify that it complies with such requirements of Parts II of these regulations as apply to the replacement by virtue of paragraph (2).

    (6) A person who makes any such replacement of a service pipe, shall ensure that, as soon as is reasonably practicable, any part of the old pipe which is not removed is disconnected as near to the main as is reasonably practicable.

EXPLANATORY NOTE
(This Note is not part of the Regulations.)

These Regulations impose requirements as to the installation and use of gas fittings for the purpose of securing that the public is so far as is practicable protected from personal injury, fire, explosion or other damage arising from the use of gas supplied through pipes. Gas fittings are gas pipes, fittings, meters, apparatus and appliances designed for use by consumers of gas (Gas Act 1972, s. 48(1)) but not service pipes except where part of a meter installation. The Regulations do no apply in respect of mines and factories (Regulation 2(3)).

Part B (Regulations 3 to 9) of the Regulations contains provisions of general application. Regulations 3(1) and 4 to 6 and 9 impose requirements on persons installing or working on gas fittings. Regulations 3(2) requires employers of such persons to ensure that Regulation 3(1) and the following provisions of the Regulations are complied with. Regulation 7 imposes requirements on persons doing things affecting gas fittings or their ventilation. Regulation 8 prohibits a new supply of gas being given to a building unless there are adequate emergency controls.

Part C (Regulations 10 to 16) of the Regulations contains provisions relating to meter installations. Regulations 11, 12(1) and (3), 13(1) and (2) and 15(1) impose requirements on installers. Regulations 13(3), 14, 15(2) and (3) may affect suppliers of gas or other persons.

Part D (Regulations 17 to 23) of the Regulations contains provisions relating to installation pipes. For the most part they impose requirements on persons installing or working on installation pipes. Regulations 21(3) and 22(2) impose requirements on suppliers of gas and those responsible for premises supplied respectively.

Part E (Regulations 24 to 34) contains provisions relating to gas appliances. For the most part they impose requirements on persons installing or working on gas appliances. Regulations 33(3), 34(1) and 34(2) impose requirements on suppliers of gas, persons using gas appliances and persons working on gas mains, service pipes and gas fittings respectively.

Part F (Regulations 35 to 38) contains miscellaneous provisions. Regulation 35 deals with escapes of gas. Regulation 36 provides for a penalty of a maximum of £2,000 on summary conviction for persons found guilty of contravening or failing to comply with the Regulations. Regulation 37 provides, in respect of certain provisions of the Regulations, a defence if the accused can show that he took all reasonable steps to prevent contravention of or failure to comply with the Regulations. Regulation 38(1) amends the Gas Safety Regulations 1972 which apply only to gas supplied through pipes by the British Gas Corporation. Part III and VI (installation of meters, installation pipes and gas appliances and use of gas respectively) cease to have effect leaving Part I (general), Part II (service pipes) and Part VII (removal, disconnection, alteration, replacement and maintenance of gas fittings etc.) in force. Part VII is amended so as to confine its operation to service pipes and their associated pipe fittings and a provision is added providing for the removal of the old pipe when a service pipe is replaced. Regulation 38(2) increases the maximum fine for contraventions of or failure to comply with the Gas Safety Regulations 1972 and the Gas Safety (Rights of Entry) Regulations 1983 from £1,000 to £2,000.

The publications referred to in Regulation 11(3) may be obtained from the British Standards Institution, 2 Park Street, London W1A 2BS.

GAS SERVICES TRAINING COURSE — APPENDIX A

B.S.I. CODES OF PRACTICE: for Installing Domestic Gas Appliances

B.S. 6400; 1985	Specification for installation of domestic gas meters (2nd family gases)
B.S. 6798; 1987	Specification for installation of gas fired hot water boilers rated input not exceeding 60kW
B.S. 5440; Part 2; 1990	Flues and air supply for gas appliances of rated input not exceeding 60kW (1st and 2nd family gases). Part 2: Air Supply.
B.S. 5440; Part 1; 1989	Flues and air supply for gas appliances of rated input not exceeding 60kW (1st and 2nd family gases). Part 1: Flues.
B.S. 5482; Part 2; 1977 (1988)	Domestic butane and propane gas burning installations: installation in caravans and non-permanent dwellings.
B.S. 5482; Part 1; 1979 (1989)	Domestic butane and propane gas burning installations: installations in permanent dwellings.
B.S. 5546; 1990	Installation of gas hot water supplies for domestic purposes.
B.S. 5864; 1989	The installation of gas fired ducted air heaters of rated input not exceeding 60kW.
B.S. 5871; 1980	Installation of gas fires, convectors and fire/back boilers (2nd family gases)
B.S. 6172; 1982	Installation of domestic gas cooking appliances. (2nd family gases).
B.S. 6714; 1986	Installation of decorative gas log and other fuel effect appliances (1st, 2nd and 3rd family gases).
B.S. 6891; 1988	Installation of low pressure gas pipework up to 28mm (R1) in domestic premises (2nd family gas).

For sets of these Codes please apply to the British Standards Institution, Linford Wood, Milton Keynes MK14 6LE.

Joyce